MintDuino

**James Floyd Kelly and
Marc de Vinck**

O'REILLY®

Beijing · Cambridge · Farnham · Köln · Sebastopol · Tokyo

MintDuino

by James Floyd Kelly and Marc de Vinck

Published by O'Reilly Media, Inc., 1005 Gravenstein Highway North, Sebastopol, CA 95472.

O'Reilly books may be purchased for educational, business, or sales promotional use. Online editions are also available for most titles (*http://my.safaribooksonline.com*). For more information, contact our corporate/institutional sales department: (800) 998-9938 or *corporate@oreilly.com*.

Editor: Brian Jepson
Production Editor: Kristen Borg
Proofreader: O'Reilly Production Services
Cover Designer: Mark Paglietti
Interior Designer: David Futato
Illustrator: Robert Romano

September 2011: First Edition.

Revision History for the First Edition:
September 12, 2011 First release
December 16, 2011 Second release

See *http://oreilly.com/catalog/errata.csp?isbn=9781449307660* for release details.

ISBN: 978-1-449-30766-0
[LSI]
1323973843

Contents

Preface

The MintDuino is deceiving—this little tin of electronics is capable of providing the brains for an unlimited number of devices limited only by your imagination...and, of course, your bank account.

While the MintDuino is certainly capable of being used during a moment of inspiration, you'll find that the real power of the device comes when you sit down, figure out what you want to make, and then start tackling your project a bit at a time. Anyone who has done circuit building or programming (or both) knows that it rarely pays to just start inserting components and writing complex sketches (programs in Arduino-speak). When trouble arises in a circuit or a program, it can be a time-consuming process to track down the error (or more likely, *errors*) when you have too much complexity. That's why it's always a good idea to break a project idea down into manageable components—let's call them subtasks. If your big project has five major subtasks, tackling the first subtask and building a slightly less complicated circuit with a slightly less complicated sketch will save time and reduce stress. Once you've got the first subtask figured out, you move on to the second. By the time you've got all five subtasks built and their individual sketches tested, you're likely to have an increased chance of success when you pull them all together and build The Big Project.

And that's exactly what you're going to do in this MintDuino Notebook. You're going to be shown The Big Project...but you're not going to be allowed to build it just yet. The Big project has subtasks, and first you're going to learn how to get the subtasks to work. Once you've done that, you'll be ready for the finale where you bring together all you've learned and tackle...The Big Project. So, what is The Big Project? Keep reading.

What You Need

The complete list of materials for this project is below, but you'll find a partial list of components with each subtask that contains only those items used in that particular section:

- 1 MintDuino—assembled (see *http://makeprojects.com/Project/Build -a-Mintronics-MintDuino/608/1* for assembly instructions)
- 1 9V battery
- 1 FTDI adapter, such as the FTDI Friend (see *http://www.makershed .com/ProductDetails.asp?ProductCode=MKAD22*)
- 1 USB cable (A to mini-B type)
- A jumper wire kit, such as *http://www.makershed.com/ProductDetails .asp?ProductCode=MKSEEED3*
- 2 LEDs (you can use any color you have handy)

Plus, you'll need the following components, all of which are available in the Mintronics: Survival Pack (*http://www.makershed.com/ProductDetails .asp?ProductCode=MSTIN2*):

- 2 LEDs (one red and one green LED come with the Survival Pack, but you can use any color you have handy)
- 1 mini breadboard
- 1 9V battery connector
- 3 resistors (100 ohm minimum)
- Jumper wires (you'll find enough wire in the Survival Pack to get you started, but you'll need to dip into the jumper wire kit soon)
- 2 pushbuttons
- Enough jumper wire to get you through Subtask 3.

Conventions Used in This Book

The following typographical conventions are used in this book:

Italic
> Indicates new terms, URLs, email addresses, filenames, and file extensions.

`Constant width`
> Used for program listings, as well as within paragraphs to refer to program elements such as variable or function names, databases, data types, environment variables, statements, and keywords.

Constant width bold

Shows commands or other text that should be typed literally by the user.

Constant width italic

Shows text that should be replaced with user-supplied values or by values determined by context.

  TIP: This icon signifies a tip, suggestion, or general note.

  CAUTION: This icon indicates a warning or caution.

Using Code Examples

This book is here to help you get your job done. In general, you may use the code in this book in your programs and documentation. You do not need to contact us for permission unless you're reproducing a significant portion of the code. For example, writing a program that uses several chunks of code from this book does not require permission. Selling or distributing a CD-ROM of examples from O'Reilly books does require permission. Answering a question by citing this book and quoting example code does not require permission. Incorporating a significant amount of example code from this book into your product's documentation does require permission.

We appreciate, but do not require, attribution. An attribution usually includes the title, author, publisher, and ISBN. For example: "*MintDuino* by James Floyd Kelly (O'Reilly). Copyright 2011 Blue Rocket Writing Services, Inc., 978-1-4493-0766-0."

If you feel your use of code examples falls outside fair use or the permission given above, feel free to contact us at *permissions@oreilly.com*.

Safari® Books Online

  Safari Books Online is an on-demand digital library that lets you easily search over 7,500 technology and creative reference books and videos to find the answers you need quickly.

With a subscription, you can read any page and watch any video from our library online. Read books on your cell phone and mobile devices. Access new titles before they are available for print, and get exclusive access to manuscripts in development and post feedback for the authors. Copy and paste code samples, organize your favorites, download chapters, bookmark key sections, create notes, print out pages, and benefit from tons of other time-saving features.

O'Reilly Media has uploaded this book to the Safari Books Online service. To have full digital access to this book and others on similar topics from O'Reilly and other publishers, sign up for free at *http://my.safaribooksonline.com*.

How to Contact Us

Please address comments and questions concerning this book to the publisher:

O'Reilly Media, Inc.
1005 Gravenstein Highway North
Sebastopol, CA 95472
800-998-9938 (in the United States or Canada)
707-829-0515 (international or local)
707-829-0104 (fax)

We have a web page for this book, where we list errata, examples, and any additional information. You can access this page at:

http://www.oreilly.com/catalog/0636920020882

To comment or ask technical questions about this book, send email to:

bookquestions@oreilly.com

For more information about our books, courses, conferences, and news, see our website at *http://www.oreilly.com*.

Find us on Facebook: *http://facebook.com/oreilly*

Follow us on Twitter: *http://twitter.com/oreillymedia*

Watch us on YouTube: *http://www.youtube.com/oreillymedia*

Content Updates

December 16, 2011

- Modified the book's style and trim size.
- Added a new chapter that explains how to build the Mintduino.

1/Build a Mintronics: MintDuino

The MintDuino is perfect for anyone interested in learning (or teaching) the fundamentals of how micro controllers work. It will have you building your own micro controller from scratch on a breadboard, and then easily programming it from almost any computer via the Arduino programming environment.

Unlike pre-built micro controllers, the MintDuino demonstrates the specific relationship between the wires, resistors, capacitors, and integrated circuits that enables you to program the micro controller from your computer. After building the MintDuino, you'll have a much better understanding of how micro controllers work, and how electronics can interact with the physical world. This chapter explains how to assemble the MintDuino; in Chapter 2, you'll learn how to create a simple game with it.

Build the Power Supply

Start building your MintDuino by adding the 7805 power regulator. This converts the 9v power to 5v power that the ATMega can use. Insert the 7805 into column "i" on the breadboard and rows 1, 2, and 3. The metal heatsink should be facing the right (or column "J"). Now we are going to add two 10 µF capacitors to the power regulator:

1. I like to trim the leads down so they don't stick so far out of the breadboard. One lead is longer than the other. The long lead is the (+) lead and the short one is the (-) lead. If you trim it, make sure to keep the lengths different lengths so it's easy to identify the (+) and (-) leads.

2. Take the first capacitor and insert the (+) lead into "g1" and the negative lead into "g2". Easy!

3. Take the other 10 µF capacitor, and insert the (-) lead into row 1 of the (-) power rail of the breadboard. Insert the (+) lead into row 1 of the (+) rail of the breadboard.

Now lets get some regulated power over to the power rails of the breadboard. Start by stripping the ends of one piece of red wire cut to approximately 1/2" long. Insert the wire from the (+) rail of the breadboard to "j3" of the bread-board. Next, strip the ends of one piece of black wire cut to approximately 1/2" long. Insert the wire from the (-) rail of the breadboard to "j2" of the breadboard. Your breadboard should look like Figure 1-1

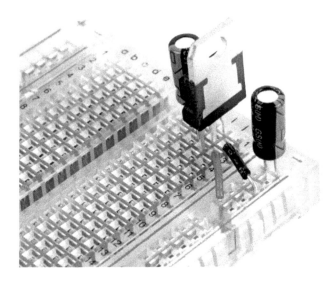

Figure 1-1. *Breadboard with voltage regulator, capacitors, and wire*

Now it's time to add the power LED. Start by cutting down the leads, just as you did on the capacitor. Make sure to keep the long one (+) longer than the short one (-)! Now you can insert the red LED into the breadboard: the longer lead (+) goes into "d2" and the negative (-) goes into "d1". Let's get the power distributed around the board and to the LED:

1. Start by cutting one red wire, approximately 1/2" long and one black wire, approximately 1/2" long. Strip both ends of each wire.

2. Insert the red wire from "f1" to "e4", and the black wire from "f2" to "e5".

3. Cut another piece of black wire about 1/2" long (from here on out, I'm going to stop reminding you to strip each end, so make sure you do it) and insert it from the (-) rail of the breadboard and "b1".

4. While we are here, lets add a 220 Ohm resistor (red, red, brown) from the (+) rail of the breadboard to "b2". This will limit the amount of current that goes into the LED, and keep it from burning out.

5. Lastly, cut (1) piece of red and black wire about 1 1/2" long and connect the right side rails together. Remember to connect (+) to (+) and (-) to (-). Your breadboard should look just like Figure 1-2

Figure 1-2. *Breadboard ready to be powered up*

Now we can power it up! Connect the battery clip's red wire (+) to "d4" and the black wire (-) to "d5". Connect a 9v battery and the red LED should light up. Your breadboard should look just like Figure 1-3.

 WARNING: If the LED doesn't light up, disconnect the battery immediately and double check the wiring. Take the circuit apart if necessary and start from the beginning. If the LED doesn't work at this point, nothing else will.

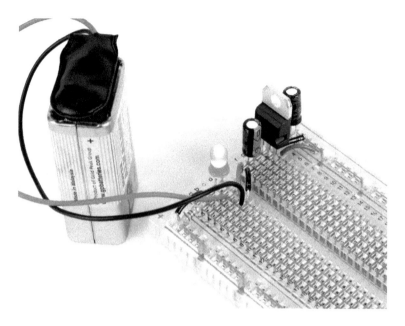

Figure 1-3. *Your breadboard is powered and ready for the next step*

Now you have a nice 5V regulated power supply from a 9V battery. Your ATMega will thank you for it! OK, enough fun. Unplug the battery and let's get started with the micro controller.

Power the Microcontroller

Now it's time to connect power to the ATMega 328 chip (also known as an integrated circuit or IC). This is the brains of your MintDuino. It combines a microprocessor, flash memory, RAM, and digital as well as analog inputs and outputs into a single chip known as a *microcontroller*. It's also the most fragile part, so make sure you've disconnected the battery before you do anything. The ATMega has a small "U" shaped notch on one end. This notch lets you know where pin 1 is on the chip. If you hold the chip vertically, with the notch on top, pin 1 is directly to the left of this notch. Insert the IC so the notch is pointing towards the power supply you just built, and so that pin 1 goes into "e9" on the breadboard.

 NOTE: You may need to bend the pins in a little bit so they don't flare out too much. Don't use a lot of force to insert the IC or you may damage the pins.

With the ATMega inserted, you should insert the 16 MHz clock crystal, which controls the speed at which the microcontroller executes instructions:

1. Insert the crystal into the breadboard at "b17" and "b18". It's not polarized, so orientation isn't important: you can insert it either way.

2. The crystal needs some capacitors to work properly. The two 22pF capacitors (marked "220") are not polarized either, so their orientation does not matter.

3. Insert one 22 pF capacitor so one pin goes into the the ground rail of the breadboard and the other into "a17".

4. Insert the other 22 pF capacitor with one pin into the the ground rail and the other into "a18".

5. While we are working on this part of the breadboard, let's connect a ground connection to the microcontroller: cut a 1/2" piece of black wire and connect the ground rail of the breadboard to "a16".

At this point, the center of your breadboard should look like Figure 1-4.

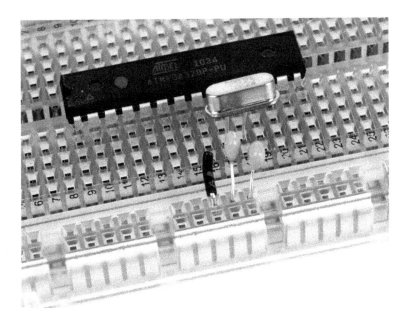

Figure 1-4. *ATMega with capacitors, crystal, and ground wire*

Now you're ready to connect the positive rail of the breadboard to the microcontroller:

1. Cut 3 pieces of red wire, all of them about 1/2" in length, and one black wire 1/2" in length.
2. Use one red wire to connect the (+) rail of the breadboard to "j16".
3. Use another piece of red wire to connect the (+) rail of the breadboard to "j17".
4. Use one black wire to connect the (-) rail of the breadboard to "j15".
5. Back to the other side of the board (by the crystal). Connect the remaining red wire from the (+) rail to "a15".

Let's wire up the green status LED, which will help us know whether everything is working properly. You can trim the LED's leads if you'd like (this will make it fit the board more snugly), but remember to keep the long lead longer than the short lead:

1. Cut one piece or red wire about 3/4" long.
2. Insert the longer lead of the LED (+) into "i24" and the shorter lead (-) into "i25".
3. Next connect the (-) ground rail of the breadboard to "j25" using a 220 Ohm resistor (red,red,brown).
4. Connect "h24" to "h18" with the red wire.

When you're done, the breadboard should look like Figure 1-5. Now we're ready for another test: connect the battery to the board as you did earlier, making sure you connect red to red and black to black. The red power LED should light up immediately, followed by the green LED. The green LED will then start blinking. This is because a simple "blink" program has already been uploaded to the ATMega. If the LEDs don't light up, immediately disconnect the power, and check all your connections again.

 NOTE: Technically, you now have an LED connected to Analog "pin 13" of the Arduino, which is the same pin used by a standard Arduino's onboard LED. However, this is not actually pin 13 of the ATMega. The Arduino development environment uses a separate pin numbering scheme (it has a set of digital pins numbered 0 through 13 and a set of analog pins numbered 0 through 5). We'll go over the pins in the last step of the build.

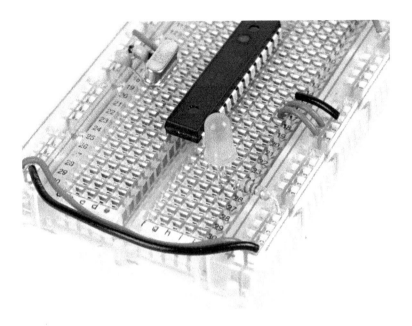

Figure 1-5. *ATMega with power connections and an LED*

Getting Ready to Program

Now you're ready to add some components that you'll need before you can program the MintDuino. Let's start with the reset button:

1. Cut one piece of black wire about 1" in length.

2. Connect the ground rail (-) of the breadboard to "d6".

3. Cut another piece of black wire about 1/2" in length and connect it from "d8" to pin 1 of the ATMega at "c9"

4. Now press the button into the breadboard. It only fits one way, so make sure the pins all line up properly (if it feels like you'll have to force it, turn it 90 degrees and try again). The four leads of the button will fit in "e6", "e8", "f6", and "f8".

5. Connect "b9" to the (+) rail of the breadboard with a 10k Ohm resistor (brown,black,orange).

You're almost there; check out Figure 1-6 to see how your breadboard should look now.

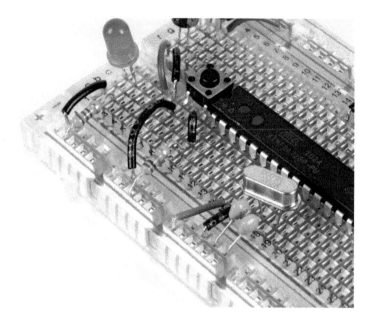

Figure 1-6. *ATMega with pushbutton*

Next, you need to wire up the six-pin programming header (only four pins are used). This doesn't connect to the microcontroller's traditional programming pins, but to Ground (-), the microcontroller's reset pin, as well as its two UART (serial port) pins, TX and RX. Unlike other microcontroller environments, Arduino is programmed over a serial connection, which is why you can use a USB to TTL serial converter such as the FTDI adapter recommended in "What You Need" on page vi.

1. Start by using a pair of pliers to adjust the pins (gently) to be centered in the plastic rail, as shown in Figure 1-7. By centering the pins it makes it much easier to plug in the FTDI adapter later.

Figure 1-7. *Centering the pins*

2. Once they are centered, insert the six-pin header in column "b" from "b25" to "b30". Now you're ready to wire up the programming pins.

3. Cut three lengths of green wire (color may vary) approximately 2" each.

4. Connect one wire from "d10" to "e27".

5. Connect the second wire from "c11" to "d26".

6. Connect the third wire from "d9" to "c23".

7. Next, add the 100 nF capacitor (marked 104 on one side and K1K on the other) from "c25" to "b23". It isn't polarized, so you can insert it either way.

8. The final step is to add a 1/2" piece of black wire from "a30" to the ground rail (-) of the breadboard. Figure 1-8 shows the completed connections.

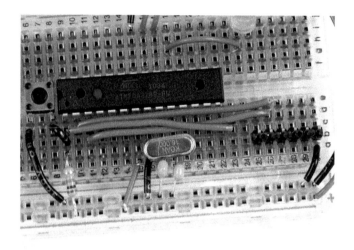

Figure 1-8. *The programming header connected to the ATMega*

You're all done! Your finished MintDuino should look like Figure 1-9. (Looks a bit like a scorpion, eh?) Now you'll be ready to program the ATMega chip when the time comes in Chapter 3.

Figure 1-9. *The finished MintDuino*

2/The MintDuino Reflex Game

The MintDuino Reflex Game will test the reflexes of two players. The game takes place on the mini breadboard, with each player waiting to push a button when an LED (the Game Light) in the center of the breadboard lights up. The Game Light will light up after a random amount of time has passed (something between, say, 5 and 10 seconds). Whichever player pushes their button first after the Game Light LED lights up wins, and a green LED will light up on the side of the winning player.

That's The Big Project—but let's break it down into four subtasks that will help us understand the proper wiring of the MintDuino, as well as the electronics components and programming elements required:

1. Wire up and program the MintDuino to light an LED—this will help us to not only light the Game Light but also the winning player's green LED.

2. Wire up and light an LED after a Random Wait Time—we will learn how to light an LED after waiting a random amount of time.

3. Wire up and program an LED to light when the pushbutton is pressed—this will help us to understand the proper wiring of a pushbutton as well as how to use it as a trigger for another event (in this case, lighting an LED).

4. Wire and program two pushbuttons to light up a matching LED when pressed—here, we'll figure out how to trigger the correct LED when its paired pushbutton is pressed.

Once we understand what's required in order to light LEDs, wait a random amount of time, and test for pushbuttons to be pressed, we can pull it all together and build the MintDuino Reflex Game. The fact that we'll have multiple LEDS shouldn't be a problem—if we know how to light one, well, we can light two...or three. And if we know how to determine if a pushbutton has been pressed, we can probably figure out how to detect which of two pushbuttons has been pressed. Then we add some code to start the game, check for a premature button push, and reset the game...and we're in business.

NOTE: This MintDuino Notebook will provide you with the sample code required to test various circuits and electronics components, but it will not be able to provide a comprehensive tutorial on programming the MintDuino (or an Arduino). If you need a better understanding of the programming language used to create MintDuino sketches, you'll want to turn to a variety of online and printed sources designed to teach beginning Arduino programming. Make: Arduino, at *http://makezine.com/arduino/*, is a great starting point. You'll find videos, projects, books, and more there.

3/Subtask 1: Light an LED

We'll start with Subtask 1 and assemble a small circuit to light a single LED. The components you will need include:

- 1 MintDuino—assembled (see Chapter 1 for assembly instructions)
- 1 9V battery
- 1 FTDI adapter, such as the FTDI Friend (see *http://www.makershed .com/ProductDetails.asp?ProductCode=MKAD22*)
- 1 USB cable (A to mini-B type)

Plus, you'll need the following components, all of which are available in the Mintronics: Survival Pack (*http://www.makershed.com/ProductDetails .asp?ProductCode=MSTIN2*):

- 1 LED (one red and one green LED come with the Survival Pack, but you can use any color you have handy)
- 1 mini breadboard
- 1 9V battery connector
- 1 resistor, 100 ohm (minimum)
- 2 jumper wires

Figure 3-1 shows the components required to assemble Subtask 1. Later, you'll use the FTDI adapter to upload the sketch to the MintDuino.

Start Building

You'll start the assembly of Subtask 1 by inserting the LED and resistor (refer to this resistor as RES1 for all subtasks) into the mini breadboard, as shown in Figure 3-1.

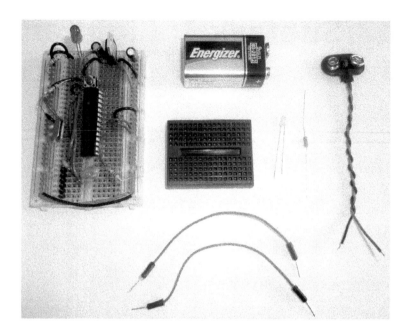

Figure 3-1. *Components required to assemble circuit for Subtask 1*

Before you continue, there are a few things you need to know when using the mini breadboard:

- The mini breadboard does not have letters or numerals to label the various rows and columns (which the MintDuino's breadboard does have).

- When wiring, rotate the mini breadboard so that it is taller than it is wider (as seen in Figure 3-2); with this orientation, each row is broken into two segments of five holes.

- The five holes in each segment share a common connection point; when inserting components, make certain that leads are inserted in different segments and not in the same grouping of five holes or the component will be shorted.

Figure 3-2. *Resistor (RES1) and LED inserted into the motherboard*

Take note that the LED has one leg that is shorter than the other. The longer leg is referred to as the *anode* or + lead (positive); and the shorter leg is referred to as the *cathode* or – lead (negative). When connecting an LED to a circuit, you must remember to connect the + lead to the voltage/supply side of a circuit and the – lead to the GND (ground) side of a circuit.

Because you haven't yet wired up power to the MintDuino, just remember (or jot down a note here) where the longer + lead of the LED is located. If you're following along with the included images, you'll want to insert the longer + lead closer to the left side of the mini breadboard; this side will be closest to the MintDuino once it is finally connected to the mini breadboard.

 NOTE: One useful way to remember how an LED is inserted into the mini breadboard is to insert the longer (+) lead closer to the MintDuino. If you consistently use this method, you'll always be able to look at an inserted LED and determine which lead is the anode and which is the cathode.

I've also inserted RES1 so that one of its leads shares a row with the cathode (−) lead of the LED. Notice in Figure 3-2 that RES1's other lead is inserted into an empty row on the mini breadboard just below the LED's anode lead.

Next, you'll use two jumper wires to connect the mini breadboard to the MintDuino. Insert one jumper wire into the same row as the LED's anode lead. Insert the other wire into the same row as RES1's non-shared lead (the lead not shared by the LED's cathode lead). This is shown in Figure 3-3. Black wire is often chosen when making connections to GND and red wire is typically selected for making connections to voltage/power; feel free to use these colors for the jumper wires if you have them (or if you have the Survival Pack), but it is not required.

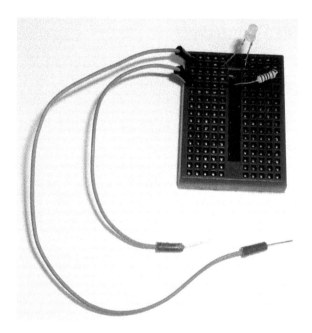

Figure 3-3. *Two jumper wires will connect the mini breadboard to the MintDuino*

Now you'll connect the two jumper wires to the MintDuino. If you've built your MintDuino based on the building instructions in Chapter 1, you'll want to connect the jumper wire connected to the LED's anode (+) lead to Pin 13 on the ATmega 328 chip. This corresponds to Row 21 on the MintDuino (again, if you've taken care to wire it up exactly as the instructions specify). You can plug that jumper wire into any free hole between *a* and *e* on Row 21.

NOTE: Pin 13 corresponds to Digital Pin 7—this information will be required shortly when we write the program to test the LED.

Plug the other jumper wire from the LED's cathode (–) into any hole on the GND column of the MintDuino. Double-check this and make absolutely certain that you've connected it to a GND column and not the PWR (5v or 3.3v) column.

Figure 3-4 shows the two jumper wires now connecting the mini breadboard to the MintDuino.

Figure 3-4. *MintDuino and mini breadboard circuit completed with jumper wires*

Upload Your First Sketch

Now it's time to upload the Subtask 1 program (sketch). You can download this sketch from *http://examples.oreilly.com/0636920020882*, or simply open your Arduino IDE and enter the following sketch/code:

```
// MintDuino NoteBook 1 - Subtask 1
int ledPin = 7;          // Digital Pin 7 for LED anode connection
int ledWaitMin = 2000;   // Set minimum wait time to 2000 milliseconds

void setup() {

    // use noise on pin 1 to generate a random number
    randomSeed(analogRead(1));

    pinMode(ledPin, OUTPUT);
}
void loop() {

    // add random time of 0-5 seconds
    int ledWait = ledWaitMin + random(5000);

    // three fast blinks
    for (int count = 0; count < 3; count++) {
        digitalWrite(ledPin, HIGH);
        delay(250);
        digitalWrite(ledPin, LOW);
        delay(250);
    }

    delay(ledWait);  // random amount of time passes
    digitalWrite(ledPin, HIGH);

    delay(2000); // wait 2 seconds after random lighting
    digitalWrite(ledPin, LOW);

    delay(5000);    // wait 5 seconds before resetting
}
```

Connect the FTDI Friend (or other FTDI adapter) to your MintDuino as seen in Figure 3-5. This connects your computer to the MintDuino so that you can upload the sketch. Remember that you'll need to provide power to the MintDuino using the 9V battery!

Figure 3-5. *Add the FTDI Friend (Adapter) to the MintDuino and upload the sketch*

After uploading the sketch to the MintDuino, leave the USB cable plugged into the FTDI/MintDuino and you should see a quickly flashing LED on the mini breadboard, as shown in Figure 3-6.

Figure 3-6. *A flashing LED lets you know that the sketch is working*

Leave the circuit wired up, as you'll use it again with Subtask 2, but unplug the USB cable that connects the FTDI/MintDuino to your computer.

Troubleshooting

If the LED is not flashing, make certain that you set the Digital Pin to 7 in the sketch and that the jumper wire from the LED anode row is connected to Pin 13 on the ATmega328 chip on the MintDuino. Also make sure that the jumper wire sharing a row with RES1 on the mini breadboard is connected to GND (ground) on the MintDuino's breadboard.

If the jumper wires are properly connected to the MintDuino, next check that you've properly inserted the LED (longer leg connected to Pin 13 via jumper wire). You might also exchange the LED for another to ensure you don't have a faulty LED.

Finally, go back and verify that the code uploaded properly to the MintDuino—if you don't see a compilation error or any error message telling you the upload failed, you can be reasonably certain that the sketch is loaded—the problem is likely a connection issue between components or a miswired circuit.

4/Subtask 2: Randomly Light an LED

Subtask 2 will use the same circuit you assembled for Subtask 1; we will simply make a change to the programming that will light up the LED after a random number of seconds has elapsed. The parts required for Subtask 2 are identical to those listed in Chapter 3.

You can download the program for Subtask 2 online at *http://examples .oreilly.com/0636920020882*, or simply enter the code below into the Arduino IDE:

```
// MintDuino NoteBook 1 - Subtask 2
int ledPin = 7;         // Digital Pin 7 for LED anode connection
int ledWaitMin = 2000;  // Set minimum wait time to 2000 milliseconds

void setup() {

    // use noise on pin 1 to generate a random number
    randomSeed(analogRead(1));

    pinMode(ledPin, OUTPUT);
}
void loop() {

    // add random time of 0-5 seconds
    int ledWait = ledWaitMin + random(5000);

    // three fast blinks
    for (int count = 0; count < 3; count++) {
        digitalWrite(ledPin, HIGH);
        delay(250);
        digitalWrite(ledPin, LOW);
        delay(250);
    }

    delay(ledWait);  // random amount of time passes
    digitalWrite(ledPin, HIGH);

    delay(2000); // wait 2 seconds after random lighting
    digitalWrite(ledPin, LOW);
```

```
      delay(5000);     // wait 5 seconds before resetting
}
```

After you upload this sketch and each time you reboot or power up the MintDuino, the following will occur once:

1. The minimum wait time for turning on the LED will be initialized with a value of 2000 milliseconds (2 seconds).

2. A random seed will be generated using ambient noise (fluorescent lights, cosmic rays, radio waves, etc.) that analog pin 1 picks up.

Next, as long as the MintDuino is running (until you turn it off or its battery dies), the following actions occur over and over:

1. The main loop of the sketch starts and a random value between 0 and 5000 milliseconds will be added to the minimum wait time to generate a value between 2000 (2 seconds) and 7000 (7 seconds).

2. A *for loop* (that sets the variable count to 0, 1, and 2, in order) comes next. I only use the variable count to make sure that the code inside the loop runs exactly three times. This causes three fast flashes of the LED, alerting game players to begin watching for the game LED to light.

3. Next, there is a random pause (based on a value between 2000 and 7000 milliseconds).

4. The LED will light and hold for two seconds.

5. The LED will turn off and remain off for a five second delay.

6. The loop begins again (back to step 1).

The Game Takes Shape

As you can see, the beginnings of the MintDuino Reflex Game are starting to appear. We have a random amount of time that will pass before the LED lights up—this is what the two players will be waiting to see before they push their buttons.

And buttons is what we need to learn how to use now. We'll tackle Subtask 3 next, and learn how to light that LED by pressing a pushbutton. Unplug the USB cable from the FTDI/MintDuino and make no changes to the wiring of the LED and RES1.

NOTE: If you managed to get the LED to light in Subtask 1, then any problems you'll likely have with Subtask 2 will exist in the program itself.

If you're not seeing the initial three quick flashes of the LED, check your code to make certain you're not using the `ledWait` variable to control the on-and-off lighting in the `for` loop (there should be two delays of 250 milliseconds in the `for` loop). If you're not getting a random wait after the three initial flashes (for example, if the delay is always the same number of seconds), make certain that you're using `ledWait` to calculate the delay before taking the pin `HIGH` right after the `for` loop.

5/Subtask 3: Light an LED with a Pushbutton

Subtask 3 will add to the same circuit you assembled for Subtasks 1 and 2 by introducing the single pushbutton component to the mini breadboard, as seen in Figure 5-1. Subtask 3 will require the following additional components beyond what you've already assembled. All of these are available in the Mintronics: Survival Pack (*http://www.makershed.com/ProductDetails .asp?ProductCode=MSTIN2*):

- 1 resistor, 100 ohm (minimum)
- 1 red jumper wire (3 inches in length)
- 3 blue jumper wires (make two of them 2 inches in length, and one of them 4 inches; they can be any color except red or black)
- 1 pushbutton

 NOTE: By the end of this Subtask, you will probably have exhausted the supply of jumper wire that came with the Survival Pack, and will need to dip into your jumper wire kit if you haven't already.

Add the pushbutton to the mini breadboard by inserting it so that the distance between the two top legs is a single hole (on a single row) and the distance between a top and a bottom leg has two holes between it (spanning four rows altogether). A close-up of this can be seen in Figure 5-2.

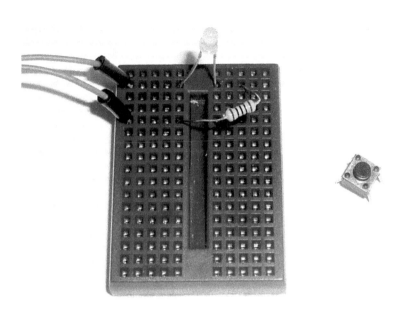

Figure 5-1. *A single pushbutton that will light the LED when pressed*

Figure 5-2. *The pushbutton is inserted into the mini breadboard in a specific way*

Figure 5-3 shows the pushbutton inserted into the mini breadboard. Make certain to leave the leftmost hole open on the top and bottom row where the pushbutton is inserted. (These will be used to add jumper wires.)

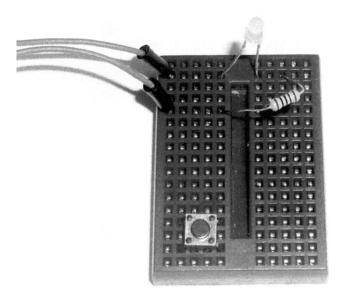

Figure 5-3. *Leave an empty hole to the left of the top row and bottom row of the pushbutton*

Connect two jumper wires to the pushbutton, as shown in Figure 5-4. One jumper wire is inserted in the first hole at the top row of the pushbutton and the second jumper wire is inserted in the empty bottom row of the mini breadboard.

Next, add a single 100 ohm resistor (this resistor will be referred to as RES2) to the mini breadboard, as shown in Figure 5-5. This resistor (RES2) will have one of its leads inserted into the row where the first resistor (RES1) is inserted, and the jumper wire goes to GND. The other lead for RES2 can be inserted in a row just above the pushbutton.

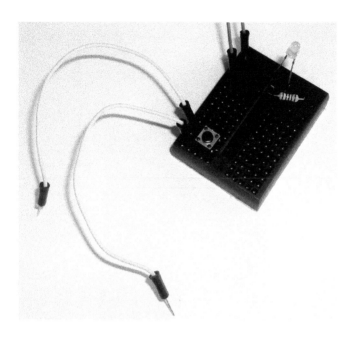

Figure 5-4. *Jumper wires will be needed to connect the pushbutton to the MintDuino*

Figure 5-5. *Add a second resistor (RES2) to serve as a pulldown resistor for the pushbutton*

As shown in Figure 5-6, insert the top jumper wire of the pushbutton into the row containing only a RES2 lead (not the row that shares GND with RES1) and insert the bottom jumper wire of the pushbutton into the bottom (empty) row of the mini breadboard (which will serve as a +5V supply for the mini breadboard).

Figure 5-6. *Wire up the pushbutton to GND and to +5V*

Finally, add two more jumper wires. Insert the red one into the bottom row of the mini breadboard and connect it to +5V on the MintDuino—we'll call this the JUMPV. Insert the other jumper wire where the pushbutton's top jumper wire and RES2 are inserted. Then connect this new jumper wire to Digital Pin 4 (or Pin 6 on the ATmega chip)—we'll call this JUMP4. This corresponds to Row 14 on the MintDuino (only if you've taken care to wire it up exactly as the online instructions specify). You can plug that jumper wire into any free hole between *a* and *e* on Row 14. Both of these new jumper wires can be seen in Figure 5-7.

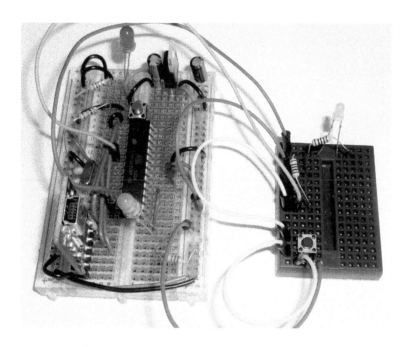

Figure 5-7. *Connect the circuit to the MintDuino*

Light the LED

Now you will program the the LED to light up after the pushbutton is pressed. You can download the program for Subtask 3 online at *http://examples .oreilly.com/0636920020882* or simply enter the code below into the Arduino IDE:

```
// MintDuino NoteBook 1 - Subtask 3

int ledPin = 7; // Digital Pin 7 for LED anode connection
int button = 4; // Use pin 4 for the button

void setup() {
    pinMode(ledPin, OUTPUT);
    digitalWrite(ledPin, LOW);
}

void loop() {
    int state = digitalRead(button);
    if (state == HIGH) { // determine if button is pressed or not
      lightLED();        // if it is, light the LED
    }
```

```
  }

  void lightLED(){ // only called when the button state is HIGH (pressed)
    digitalWrite(ledPin, HIGH);
    delay(1000);
    digitalWrite(ledPin, LOW);
  }
```

Upload the sketch to the MintDuino and the following will occur:

1. The LED will not light (at first) as it waits for the button to be pressed.

2. The program will loop forever, waiting for the button to be pressed.

3. When you press the button, the `state` variable is set to `HIGH`.

4. If `state` is `HIGH`, the `lightLED` function is called.

5. When the `lightLED` function is called, the LED stays lit for 1 second and then turns off.

6. The program waits for the button to be pressed again.

Now we're getting close. We know how to light an LED, create random time before the LED lights up, and we know how to wire up a button to light up the LED. But to make this a real game, there must be at least two opponents, and that means adding a second button to the mix...which is exactly what we'll do in Subtask 4.

NOTE: If the LED is not lighting up when you press the pushbutton, check your wiring. The pushbutton must be connected to +5 volts—check that a jumper wire (JUMPV) is connecting the pushbutton to the bottom row of the mini breadboard that is supplying the voltage.

Also verify that the other jumper wire leaving the pushbutton shares a row with RES2 and JUMP4 on one end, and is inserted into pin 6 on the ATmega chip on the other end.

6/Subtask 4: Add Buttons and LEDs

Subtask 4 will require the following components beyond what you've already used in the earlier subtasks:

- 1 resistor (100 ohm minimum), included in the Survival Pack
- 5 jumper wires (by now, you have probably used up the jumper wire in the Survival Pack, so you'll need to dip into the jumper wire kit recommended in Chapter 2)
- 1 pushbutton, included in the Survival Pack
- 2 LEDs (preferably green; the Survival Pack includes 1 green LED)

Subtask 4 will add to the circuit you assembled for Subtask 3 by introducing a second pushbutton component and two new LEDs to the mini breadboard. You can see these components in Figure 6-1.

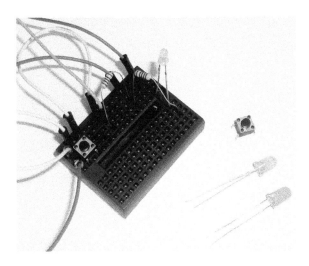

Figure 6-1. *An additional pushbutton and two LEDs will be added*

Let's build it. Insert the second pushbutton (we'll call it PUSH2; we'll call the one that's already in there PUSH1) on the opposite side of the mini breadboard, as shown in Figure 6-2.

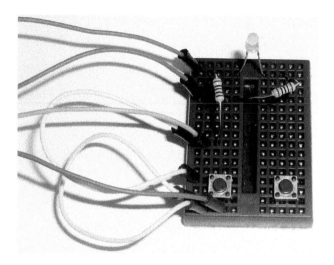

Figure 6-2. *Insert the second pushbutton on the opposite side of the mini breadboard*

Now we'll insert one LED for each player. Each LED will be placed above a pushbutton as shown in Figure 6-3. Insert each LED so that the long lead (anode) is on the row just above the pushbutton and the short lead (cathode) is on the row above that.

Now use two jumper wires to connect the cathode (short) leads from the two new LEDs to GND via pulldown resistors. The jumper wire for the Player 1 LED's cathode (on the left, closest to the MintDuino) should go to RES2, the pulldown resistor for Button 1. The jumper wire for the Player 2 LED will go to a new 100 ohm pulldown resistor (referred to as RES3) for Button 2. Insert one RES3 lead into an empty row on the mini breadboard; the other lead will go to GND where both RES1 and RES2 are connected. This can all be seen in Figure 6-4.

Next you'll connect the anode (long) lead from each LED to the ATmega chip. Connect the Player 1 LED (on the left side of the mini breadboard) to the ATmega chip with a jumper wire going to Digital Pin 5 (pin 11 on the ATmega chip). This corresponds to Row 19 on the MintDuino (only if you've taken care to wire it up exactly as the online instructions specify). You can plug that jumper wire into any free hole between *a* and *e* on Row 19.

Figure 6-3. *Each player will have an LED that corresponds to a pushbutton*

Figure 6-4. *Connect new LEDs to GND with a pulldown resistor*

Now, connect the Player 2 LED to the ATmega chip with a jumper wire going to Digital Pin 6 (pin 12 on the ATmega chip). Figure 6-5 shows the two new jumper wires connecting the LEDs to the MintDuino. This corresponds to Row 20 on the MintDuino (again, only if it's wired up exactly written as in the online instructions). Plug that jumper wire into any free hole between *a* and *e* on Row 20.

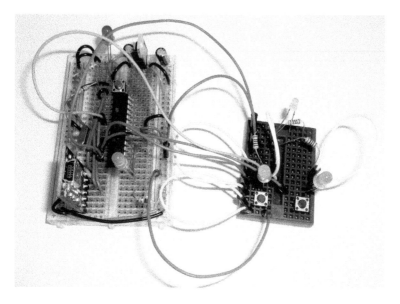

Figure 6-5. *Connect LEDs to the MintDuino with jumper wires*

 NOTE: The jumper wires are starting to crowd the mini breadboard at this point, so feel free to move resistors, jumper wires, and LEDs to other locations if you need to do so. If necessary, you can also replace the flexible jumper wires with shorter connector wires to declutter.

We now need to finish wiring up the Player 2 button (the Player 1 button keeps the same wiring from Subtask 3). Use a jumper wire to connect the bottom lead of PUSH2 to the +5V row on the bottom of the mini breadboard. Next, connect a jumper wire from the top lead of PUSH 2 to the RES3 you added for the Player 2 LED. Finally, add a jumper wire (we'll call it JUMP5) that connects RES3 (for the Player 2 LED) to Digital Pin 3 (pin 5 on the ATmega chip), which corresponds to row 13 on the MintDuino breadboard. Figure 6-6 shows the circuit for Subtask 4 wired up and ready for its program.

Figure 6-6. *Connect the Player 2 pushbutton to the MintDuino*

Light LEDs for Each Player

Now it's time to program the MintDuino so that a player's LED will light up when that player's pushbutton is pressed. You can download the program for Subtask 4 at *http://examples.oreilly.com/0636920020882* or simply enter the code below into the Arduino IDE:

```
// MintDuino NoteBook 1 - Subtask 4

int ledPin = 7;  // Digital Pin 7 for LED anode connection
int ledPlayer1 = 5;
int ledPlayer2 = 6;
int button1 = 4;
int button2 = 3;

void setup() {
  pinMode(ledPin, OUTPUT);
  pinMode(ledPlayer1, OUTPUT);
  pinMode(ledPlayer2, OUTPUT);
}
void loop() {
```

```
int state1 = digitalRead(button1);
if (state1 == HIGH) {  // determine when button is pressed
  lightLED1(); // if button is pressed, call the lightLED function
}


int state2 = digitalRead(button2);
if (state2 == HIGH) {
  lightLED2();
}
}

void lightLED1(){  // only called when the button state is HIGH (pressed)
  digitalWrite(ledPlayer1, HIGH);
  delay(1000);
  digitalWrite(ledPlayer1, LOW);

}

void lightLED2(){  // only called when the button state is HIGH (pressed)
  digitalWrite(ledPlayer2, HIGH);
  delay(1000);
  digitalWrite(ledPlayer2, LOW);
}
```

Upload the sketch to the MintDuino and the following will occur:

1. Neither LED will light (at first) as each waits for its respective button to be pressed.

2. The program will loop forever, waiting for a button to be pressed.

3. When the Player 1 button is pressed, the state1 variable is set to HIGH.

4. When the Player 2 button is pressed, the state2 variable is set to HIGH.

5. If state1 is HIGH, the lightLED1 function is called.

6. If state2 is HIGH, the lightLED2 function is called.

7. When the lightLED1 function is called, the Player 1 LED stays lit for 1 second and then turns off.

8. When the lightLED2 function is called, the Player 2 LED stays lit for 1 second and then turns off.

9. The program waits for a button to be pressed again.

You've probably figured out that the circuit for the MintDuino Reflex Game is done. If Subtask 4 is working properly for you, then you've got all the buttons, LEDs, and resistors wired up correctly. All that's left is to write the new sketch for the game.

NOTE: If one of the player LEDs is not lighting up when you press its corresponding pushbutton, check to see that the pushbutton is connected to +5 volts and that a jumper wire is connecting that pushbutton to the bottom row of the mini bread-board that is supplying the voltage. Check each pushbutton to verify that it is connected to the proper Digital Pin on the ATmega chip by a jumper wire that shares a pulldown resistor going to GND. Finally, check the orientation of the player LEDs to make certain the anode and cathode leads are inserted properly—the anode leads will go to pins on the MintDuino and the cathode leads will go to pulldown resistors connected to GND.

7/Subtask 5: Program the Game

Now that you've got the mini breadboard circuit wired up properly and connected to the MintDuino, all that's left is to upload the sketch that will allow two players to see who is the fastest button pusher on the planet.

The game will run as follows:

1. Turn on the MintDuino—use either a 9V battery or USB power via the FTDI Adapter.
2. Both player LEDs will light up and stay lit.
3. Press the Player 2 button to start the game; both player LEDs will turn off.
4. The Game Light (center LED on the mini breadboard) will blink three times.
5. After the Game Light blinks three times, a random amount of time will pass before it lights again.
6. When the Game Light blinks again, each player will try to push his or her button before the other player.
7. The fastest player's LED will light up to indicate the winner.
8. Pressing a button before the delay is over will not light an LED.
9. Press the Player 2 button to start a new game.

Figure 7-1 shows the diagram of the final circuit (all resistors are 100 ohm). The diagram was created using Fritzing, an open source tool for designing interactive electronics. See *http://fritzing.org*.

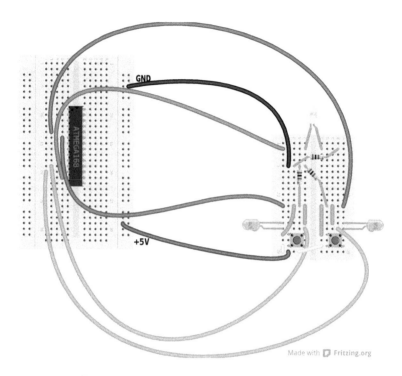

Figure 7-1. *Circuit diagram for the MintDuino Reflex Game*

The Final Sketch

You can download the program for the MintDuino Reflex Game at *http://examples.oreilly.com/0636920020882* or simply enter the code below into the Arduino IDE:

```
// MintDuino NoteBook 1 - Reflex Game version 7.0
int ledGameLight = 7;  //Digital Pin 7 for LED anode connection
int ledPlayer1 = 5;    //Digital Pin 1 for Player 1 LED
int ledPlayer2 = 6;    //Digital Pin 2 for Player 2 LED
int button1 = 4;       //Digital Pin 4 for Player 1 button
int button2 = 3;       //Digital Pin 5 for Player 2 button
int state2 = 0;
int state1 = 0;
int ledWait = 5000;    //Wait time will be a minimum of 5 seconds

void setup() {
  pinMode(ledGameLight, OUTPUT);
  pinMode(ledPlayer1, OUTPUT);
  pinMode(ledPlayer2, OUTPUT);
```

```
  pinMode(button1, INPUT); // is this needed?
  pinMode(button2, INPUT); // is this needed?
  randomSeed(analogRead(1)); //use Analog Pin 1 to generate a random number
}

void loop(){
  state2 = digitalRead(button2);    // Read the state of the pushbutton value

  if (state2 == HIGH) {      // Check if the pushbutton is pressed
    digitalWrite(ledPlayer1, LOW);    // Turn LED off:
    digitalWrite(ledPlayer2, LOW);    // Turn LED off:
    delay (2000);
    beginGame();

  }
  else {
    // turn LED off:
    digitalWrite(ledPlayer1, HIGH);    // Turn LED on
    digitalWrite(ledPlayer2, HIGH);    // Turn LED on
  }
}
void beginGame(){      // Only called when the button state is HIGH (pressed)

  // three fast blinks
  for (int count = 0; count < 3; count++) {
    digitalWrite(ledGameLight, HIGH);
    delay(500);
    digitalWrite(ledGameLight, LOW);
    delay(500);
  }

  // Now generate a wait time before Game Light turns on
  ledWait = 5000;  //reset value to minimum of 5 seconds
  ledWait = ledWait + random(5000); // add random value 0-5000 milliseconds

  //Turn on Game Light after Wait Time expires
  delay(ledWait);
  digitalWrite(ledGameLight, HIGH);
  delay(100);
  digitalWrite(ledGameLight, LOW);

  int gameOver = 0;

  while (!gameOver) {
    //determine which player button was pressed first
    int button1State = digitalRead(button1);
    int button2State = digitalRead(button2);

    if (button1State != button2State) {
```

```
      delay(5); // pause, then take another reading
      if (button1State == HIGH && digitalRead(button1) == HIGH) {
        Player1Win();
        gameOver = 1;
      }

      if (button2State == HIGH && digitalRead(button2) == HIGH) {
        Player2Win();
        gameOver = 1;
      }
    }
    else {
      if (button1State == HIGH && button2State == HIGH) {
        // tie
        itsATie();
        gameOver = 1;
      }
    }

  }

  // Start game over

}

// Tie
void itsATie() {

  for (int i = 0; i < 3; i++) {
    digitalWrite(ledPlayer1, HIGH);
    digitalWrite(ledPlayer2, HIGH);
    delay(250);
    digitalWrite(ledPlayer1, LOW);
    digitalWrite(ledPlayer2, LOW);
    delay(250);
  }
}

// Player 1 won, light his/her LED
//
void Player1Win() {
  digitalWrite(ledPlayer1, HIGH);
  digitalWrite(ledPlayer2, LOW);
  delay(4000);
  digitalWrite(ledPlayer1, LOW);
}

// Player 2 won, light his/her LED
//
void Player2Win() {
```

```
        digitalWrite(ledPlayer2, HIGH);
        digitalWrite(ledPlayer1, LOW);
        delay(4000);
        digitalWrite(ledPlayer2, LOW);
    }
```

After uploading the sketch to the MintDuino, run the program. When the game first starts, the two Player LEDs will light up and stay lit until Player 2 presses his button. The Player LEDs will turn off and the Game Light will blink three times. After the third blink, both players will wait until the Game Light blinks a fourth time and then try to be the first player to press their button. The player that wins will have her LED light up for four seconds before turning off. Press the Player 2 button to play another game.

 NOTE: If you were able to successfully run Subtask 4, then any errors you encounter are most likely in the sketch for the game. Make certain you've uploaded the correct sketch, and check that all of your jumper wires are inserted properly in the mini breadboard and the MintDuino. Check that both pushbuttons are pushed securely into the mini breadboard and make certain that all LEDs are wired up correctly with respect to their anode and cathode leads.

Conclusion

There's no arguing that the MintDuino Reflex Game is relatively simple in terms of gameplay. But think about what you've done for a moment—with two pushbuttons, three LEDs, three resistors, and a handful of jumper wires, you transformed the MintDuino and mini breadboard into a functional game. It's not complex and it's certainly not pretty to look at, but it works!

Hopefully you're starting to see the power that resides in that small MintDuino tin and maybe pondering some of your own special projects. If you're not yet ready to leave the MintDuino Reflex Game and want to dive a little deeper, here are some suggestions on ways to improve the game:

Turn it into a 3- or 4-player game
    It might require a slightly larger breadboard, but since you already know how a 2-player game works, it's not a big jump to program it for additional pushbuttons and LEDs.

Keep score

Consider adding to the sketch the required code to cause the MintDuino's status light to blink the Player 1 score, pause three seconds, and then blink the Player 2 score. Or you could add another LED for each player that flashes their score before a new game begins.

Penalize early button pushing

Modify the sketch to check to see if a player pushes his or her button before the Game Light turns on. If the button is pressed before the Game Light blinks, that player automatically loses and the other player's LED turns on.

Move the game to a project box

With a perf-board and a project box, you could easily give the game a permanent home. With a project box, you could even add an LCD screen to display the score or maybe a speaker that plays a series of beeps instead of lighting the Game LED.

Congratulations on completing the MintDuino Reflex Game! If you're looking for more projects for your MintDuino, be sure to check out the Make: Projects website at www.makeprojects.com (*http://www.makeprojects.com*) and look over all the Arduino projects for something that catches your eye. Your MintDuino might not look like the Arduinos you see online, but it's got the same functionality and can easily be substituted. You can also find many more projects on the various Arduino-related websites, including the official Arduino site at www.arduino.org (*http://www.arduino.org*), the community-run www.arduino.cc (*http://www.arduino.cc*), and instructables.com (*http://instructables.com*).

Special thanks to Brian Jepson and Will Price for their assistance with the MintDuino Reflex Game sketch.

About the Authors

Jim Kelly was accepted into the LEGO MINDSTORMS Developer Program (MDP) in early 2006 and helped to beta test the LEGO MINDSTORMS NXT kit and software. He is a member of the MINDSTORMS Community Partners, a group that continues to assist LEGO with testing and growing the NXT product.

Marc de Vinck is the Director of Product Development at MAKE and a member of the MAKE Technical Advisory board. He's worked in several different fields including a period of time as a traditional metalsmith, Illustrator, and 3D model maker.